Patrick Kimuyu

AF150845

Literature Review on the Application of Nanotechnology in Tissue Engineering

GRIN Publishing

GRIN - Your knowledge has value

Since its foundation in 1998, GRIN has specialized in publishing academic texts by students, college teachers and other academics as e-book and printed book. The website www.grin.com is an ideal platform for presenting term papers, final papers, scientific essays, dissertations and specialist books.

Visit us on the internet:

http://www.grin.com/

http://www.facebook.com/grincom

http://www.twitter.com/grin_com

THE USE OF NANOBIOTECHNOLOGY IN TISSUE ENGINEERING

PATRICK KIMUYU

Introduction

Over the past few decades, the field of tissue engineering seems to have been receiving extensive attention due to its rapid growth. The motives behind the extensive focus on tissue engineering, primarily the use of nanobiotechnology in tissue engineering are based on its capability to offer treatment alternatives through restoring the biological function and the anatomic structure of an injured, missing, or damaged tissue or organ which occur as a result of a pathological process or injury (Kingsley, Ranjan, Dasgupta & Saha, 2013; Wen, Shi & Zhang, 2005). This novel approach to treatment combines the use of nanomaterials with cells to facilitate regeneration of tissues or organs. As such, the use of nanotechnology has revolutionized tissue engineering, and provided new treatment interventions which hold the promise of saving the lives of several millions of patients around the globe (Markovic et al., 2016). Nanobiotechnology has solved the existing challenge in tissue engineering using the contemporary therapies including poor vascularization of cells, low anatomical integrity of engineered cells/or tissues, immunological incompatibility with the host, and lack of functional cells (Rivron et al., 2008). Therefore, this paper provides a systematic review of literature on the application of nanotechnology in tissue engineering.

Literature Review

Nanomaterials for Tissue Engineering

Through nanotechnology, several nanomaterials have been created for use in tissue engineering. These materials fall into different structures including nanopatterns, nanofibers and controlled-release nanoparticles. From a biological perspective, these nanomaterials are designed to mimic native tissues because they are engineered to be of nanometer size similar to extracellular fluid and other cellular components (Guen, Lifeng & Ali, 2007), in order to facilitate regeneration. In this context, three nanomaterials for tissue engineering are discussed.

Self-assembled nanomaterials

These materials are synthesized through the use of electrolytic deposition, PH induction and biometric coating to induce self assembly. Some of the biomolecules used in preparing self-assembled nanomaterials for tissue engineering include chitosan, peptide amphiphile, amelogenin/apatite, and hyaluronan (Guen, Lifeng & Ali, 2007). In principle, the formation of self-assembled peptides is facilitated by the existence of hydrophilic and hydrophobic regions of the peptides combined with charge shielding by the use of hydrogels (Hartgerink, Beniash & Stupp, 2002). Recently, 3D self-assembled nanofibers have been developed through the use of peptide amphiphile. These fibers are used to aid bone morphogenetic protein-protein attachment through the conjugation of peptide amphiphile with arginine-glyccine-aspartate acid. Additionally, electrolytic deposition is employed in synthesizing nanomaterials that enable collagen fibers to grow at cathode, primarily in bio-compositing enamels in dental therapy and osteotherapy. Moreover, electrolytic deposition is used to coat self-assembled calcium phosphate and amelogenin (Wang, Apeldoorn & Groot, 2005).

Electrospun nanofibers

Nanofibers are used to develop biomimic scaffold for tissue engineering, primarily in cardiac and bone tissues. In practice, nanofibers have been found to be reliable materials for developing blood vessel-like structures, as well as acting as a guide for cell orientation. Ma, Kotaki, Inai and Ramakrishna (2005) report that electrospun nanonfibers have been found to control scaffold function, as well as release of encapsulated biomolecules during tissue engineering.

Nanotextured substrates

It is reported that the body contains natural nanotextures. These nanotextures influence cellular behavior in different ways. Therefore, nanotextured substrates have been synthesized to allow the manipulation of cellular behavior during tissue engineering. For instance, treating poly (lactic-co-glycolide) nanosurface with NaOH has been found to increase cell density (Miller, Thapa, Haberstroh & Webster, 2004). In principle, nanotextured substrates are meant to enhance cell vascularization and adhesion using nanotubes.

Benefits of Using Nanomaterials for Tissue Engineering

In retrospect, it is apparent that nanotechnology has immense benefits for tissue engineering. Foremost, nanofabrication techniques are useful in fabrication of biomimetic scaffold for tissue engineering. Additionally, nanomaterials have been found to create a favorable extracellular microenvironment that is essential for cell repair and replacement in damaged tissues and organs (Markovic et al., 2016). It is also reported that nanomaterials are associated with more efficiency in positioning and interaction of cells (Kingsley, Ranjan, Dasgupta & Saha, 2013). For instance, 3D culture systems exhibit an array of benefits. Some of the core benefits that are associated with nanostructures include their ability to promote the growth of primary cells and high reproducibility (Dmitriev et al., 2015). Finally, it is noted that these nanomaterials promote cell migration, as essential aspect in tissue engineering.

Applications of Nanotechnology in Tissue Engineering

Over the past few decades, the use of nanotechnology in tissue engineering has gained an unprecedented popularity. This phenomenon can be attributable to the benefits of this novel technology, especially its capability to provide advanced treatment therapies and new techniques of studying the behavior of cells. However, nanotechnology has been exploited in regenerative medicine for tissue engineering of nueral, vascular, cartilage, and bone cells. It has also been used in stem cell tissue engineering. Therefore, nanotechnology has gained a wide application in tissue engineering.

Neural cells tissue engineering

The engineering of neural cells has been enhanced by nanotechnology. In neural cell engineering, there are several nanotechniques which have been employed to achieve success in engineering neural cells. Based on the available literature, it is apparent that the engineering of neural cells employ three main nanotechniques; replica moulding, electrospinning and microcontact printing. The technique of replica moulding has been adopted to enhance the maintenance of cell behavior, as well as cell shape. In one experimental study using animal model that was carried out by Bettinger et al. (2006), nanomaterials were found to enhance the culturing of neural cells. These investigators used poly (glycerol-sebacate) microfabricated silicon to investigate the cell properties and behavior of bovine aortic endothelial cells. The findings of this study indicated that replica moulding enhances the culturing of neural cells through maintaining cell behavior and shape. In another study, the use of electrospun nanofibers was found to enhance engineering of neural cells. Xie et al. (2008) cultured neurons, astrocytes and oligodendrocytes using electrospun nanofibers, primarily polycarprolactone, and reported an improvement in cell orientation and differentiation. The results of this study were consistent with those of Yang et al. (2005) which indicated that neural stem cells exhibits increased differentiation when cultured on poly (L- lactic acid) nanofibers.

On the other hand, microcontact printing using nanofibers was found to be useful in neural cells engineering. In one study, this nanotechnique was identified to enhance the formation of synaptic connections between neural cells on polydimethylsiloxane-based protocol (Vogt et al., 2005). Finally, Schwartz and DeSimone (2008) cultured neural cells in silicon oxide and documented an increase in their action potential within one day following culturing.

Vascular cells tissue engineering

Nanotechnology has also gained an immense application in tissue engineering of vascular cells. Kurpinski et al. (2006) used polydimethylsiloxane to culture mesenchymal stem cells through the use of soft lithography and documented several beneficial effects on vascular cells. The findings of this study revealed that this technique plays integral roles in altering cell signaling, as well as inducing gene expression. In another study that was carried out by Daxini et al. (2006) culturing vascular cells with poly-urethane increases cellular retention. This property is considered essential during the implantation of vascular cells due to its capability of reducing thrombogenicity. Similar results were obtained by Lin and Helmke (2008) who applied microcontact printing technique to investigate the behavior of vascular cells when cultured with nanofibers. In this study, it was found out that culturing vascular cells with polydimethylsiloxane enhances their response to shear stress. In another study that applied electrospun nanofibers indicated that culturing vascular cells with poly (L-lactid-co- ε-caprolactone) enhances cell migration. It was also noted that electrospun nanofibers promotes the attachment of vascular cells (Xu, Inai, Kotaki & Ramakrishna, 2004). Finally, microfluidic patterning with nanofibers has been found to promote the capacity of cardiomyocytes to establish cell-ligand attachment. Culturing endothelial cells with nanofibers is reported to promote their spatial distribution (Khademhosseini, Burdick & Langer, 2004).

Stem cells tissue engineering

Nanotechnology has also found extensive application in stem cells tissue engineering. Nanofibers have been used for the delivery of hematopoietic stem cells in the bone marrow. They have also been used to improve the adhesion of stem cells during tissue engineering through the application of electrospinning technique. On the other hand, soft lithography has been used to initiate differentiation, proliferation, easing growth, and regulating the distribution of mesenchymal stem cells (Karp et al., 2006). Finally, nanofibers have been used to maintain cell integrity and orientation during tissue engineering. In one study that was carried out by Chaubey et al. (2008), nanofibers were found to enhance the differentiation of stem cells into adipocytes through the application of photolithography.

Cartilage cells tissue engineering

Biomimicked scaffolds have also been developed for culturing cartilage cells in micropatterned agarose gel. These nanostructures are considered to be useful in tissue engineering of cartilage cells in which they are used to enhance the maintenance of chondrogenic phenotype, especially through the application of photolithography technique. Additionally, nanofibers have been used to maintain controlled microenvironment for the development of cartilage cells (Chao et al., 2005).

Bone cells tissue engineering

In bone cells tissue engineering, nanotechnology has become one of the emerging novel technologies which have advanced the scope of bone tissue engineering. This technique has been investigated and adopted in various ways. In one study that was carried out by Ber, Torun and Hasirci (2005), the use of nanofibers was found to promote bone formation. Investigators in this study cultured mesenchymal osteoprogenitor cells on collagen through the application of soft lithography and noted changes in cell behavior, including the maintenance of surface topography and improved cell orientation. In another study, polycaprolactone nanofibers were found to enhance culturing of human osteogenic sarcoma cells through increasing their activity and cell morphology (Tuzlakoglu et al., 2005). These findings were consistent with those of Karp et al. (2006) who investigated the influence of nanofibers on cellular response using photocrosslinkable chisotan. This study indicated that nanofibers enhance osteospecific function of osteogenic sarcoma cells and cell adhesion. They were also found to improve groove topography for osteoblasts. Similar results were obtained by Kenar et al. (2008) who used poly (3-hydroxybutyrate-co-3-hydroxyvalerate) to culture osteoblasts through microcontact printing. Results of this study showed that culturing osteoblasts on nanofibers enhances their alignment and osteoblast adhesion during the culturing process.

Hepatic cells tissue engineering

Over the past decades, advanced techniques for regeneration of hepatic tissue have emerged. From a molecular perspective, any technology that enhances the integrity and functioning of hepatic tissue improve the ability of the body to regulate vital organ systems and processes. Therefore, the discovery of nanotechniques which are applicable in hepatic tissue engineering has opened another frontier in regenerative medicine, primarily in liver repair following pathology or hepatic injury. In animal models, nanofibers have been found to play integral roles in the development of hepatic cells. For instance, nanofibers have been found to promote the establishment of rat hepatocytes through the use of electrospinning (Xu, Inai, Kotaki & Ramakrishna, 2004). In another study that was carried out by Carraro et al. (2008), culturing rat hepatocytes with polycarbonate and polydimethylsiloxane through soft lithography was found to maintain viability of the cells. This is attributable to the fact that these nanostructures promote nutrient transfer and oxygen supply in the culture. Additionally, the use of poly (ethylene glycol) and polydimethylsiloxane in culturing hepatocytes has been found to improve cell-cell 3D orientation, as well as phenotypic functions (Khetani & Bhatia, 2008).

Gene engineering

Tissue engineering can be performed through gene engineering to synthesize tissues and cells. Currently, a nanosized optic probe has been developed to facilitate cellular growth through gene manipulations. This nanostructure is used in targeting specific genes and cellular components through which nanoparticles are used to turn genes on or off, leading to a controlled tissue growth. Additionally, nanotechnology has been applied to control plasmid penetration into cells (Wen, Shi & Zhang, 2005).

Conclusion

Conclusively, nanotechnology has advanced tissue engineering to the next frontier. As such, it is apparent that that some of the most significant challenges which have been hindering the success of regenerative medicine have been solved through the development of nanotechnology approaches. For instance, migration, cell adhesion and cell differentiation have been enhanced through the use of nanofibers. Self-assembled nanomaterials, electrospun nanofibers and nanotextured substrates have facilitated tissue engineering. Currently, an array of tissues can be engineered using nanostructured platforms. Some of the tissue cells which can be engineered through this novel technique include hepatic cells, cartilage cells, neural cells, osteoblasts, vascular cells, and stem cells. Overall, nanotechnology has provided advanced treatment therapies, as well as new approaches for studying the behavior of cells.

Future challenges and solutions

From a personal opinion, it is apparent that nanotechnology holds the promise of revolutionizing tissue engineering. These techniques are highly reproducible and easy to implement. However, the application of nanotechnology in tissue engineering faces some significant barriers which need extensive focus in future advances. For instance, it is apparent that the available composite scaffolds used for tissue engineering do not adequately satisfy the natural micro- and macrostructural characteristics of cellular physiology and biochemical dynamics.

On the other hand, some of the currently designed nanostrategies have not proved to be associated with clinical success in contemporary medicine despite their potential in animal models (Mohamed & Xing, 2012). Therefore, there is need for extensive research to identify the clinical significance of these nano-based approaches, in order to enhance the success of nanomedicine through the adoption of the new treatment therapies in improving the quality of life of millions of patients across the globe.

Ultimate success will only be achieved when nanotechnology will lead to the creation of a fully functional human organ. This will mark the transformation of the classical tissue engineering into regenerative engineering, an avenue that will provide an amicable solution to regenerative disorders which continue to cause excruciating pain to humans.

Finally, there is need to develop new strategies for reducing the cost of the available nano-based scaffolds. Affordability or rather cost is one of the key determinants of health. This implies that the high cost of nano-based therapies hinder its range of benefits to the global population. Some of the scaffolds available in the market which are associated with high cost include matrigel, ECM gel, MaxGel, PuraMatrix, and HydroMatrix (Alaribe, Manoto & Motaung, 2016). Reducing the cost of nanotherapies will increase access to millions of patients who need regenerative therapies.

References

Alaribe, F., Manoto, S., & Motaung, S. (2016). Scaffolds from biomaterials: advantages and limitations in bone and tissue engineering. *Biologia, 71*(4), 353—366. doi: 10.1515/biolog-2016-0056

Ber, S., Torun, G., & Hasirci, V. (2005). Bone tissue engineering on patterned collagen films: an in vitro study. *Biomaterials, 26*(14), 1977-1986.

Bettinger, C. J., Orrick, B., & Misra, A. (2006). Microfabrication of poly glycerolesebacate) for contact guidance applications. *Biomaterials, 27*, 2558-2565.

Carraro, A., Hsu, W., & Kulig, K. (2008). In vitro analysis of a hepatic device with intrinsic microvascular-based channels. *Biomed Microdevices, 10*(6), 795-805.

Chao, P., Tang, Z., & Angelini, E. (2005). Dynamic osmotic loading of chondrocytes using a novel microfluidic device. *J Biomech., 38*(6), 1273-1281.

Chaubey, A., Ross, K., & Leadbetter, R. (2008). Surface patterning: tool to modulate stem cell differentiation in an adipose system. *J Biomed Mater Res Part B Appl Biomater., 84B* (1), 70-78.

Daxini, S., Nichol, J., & Sieminski, A. (2006). Micropatterned polymer surfaces improve retention of endothelial cells exposed to flow-induced shear stress. *Biorheology, 43*(1), 45-55.

Dmitriev, R.I., Borisov, S.M., Kondrashina, A.V., Pakan, J.M., Anilkumar U., Prehn J.,...Zhdanov A. (2015). Imaging oxygen inneural cell and tissue models by means of anionic cell per-meable phosphorescent nanoparticles. *Cell. Mol. Life Sci., 72*, 367–381.

Geun, C., Lifeng, K., & Ali, K. (2007). Micro-and nanoscale technologies for tissue engineeringand drug discovery applications. *Expert Opin Drug Discov., 2*(12), 1-16.

Hartgerink, J., Beniash, E., & Stupp, S. (2002). Peptide-amphiphilenanofibers: a versatile scaffold for the preparation ofself-assembling materials. *Proc Natl Acad Sci USA, 99*(8), 5133-5138.

Karp, J., Yeo, Y., & Geng, W. (2006). A photolithographic method to create cellular micropatterns. *Biomaterials, 27*(27), 4755-4764.

Kenar, H., Kocabas, A., & Aydinli, A. (2008). Chemical and topographical modification of PHBV surface to promote osteoblast alignment and confinement. *J Biomed Mater Res Part A, 85A* (4), 1001-1010.

Khademhosseini, A., Burdick, A., & Langer, R. (2004). Fabrication of gradient hydrogels using a microfluidics/photopolymerization process. *Langmuir, 20*(13), 5153-5156.

Khetani, S., & Bhatia, S. (2008). Microscale culture of human liver cells for drug development. *Nat Biotechnol., 26*, 120-126.

Kingsley, J., Ranjan, S., Dasgupta, N., & Saha, P. (2013). Nanotechnology for tissue engineering: Need, techniques and applications. *Journal of Pharmacy Research, 2013*, 1-5.

Kurpinski, K., Chu, J., & Hashi, C. (2006). Anisotropic mechanosensing by mesenchymal stem cells. *Proc Natl Acad Sci USA, 103*(44), 16095-16100.

Lin, X., & Helmke, B. P. (2008). Micropatterned structural control suppresses mechanotaxis of endothelial cells. *Biophys J., 95*(6), 3066-3078.

Ma, Z., Kotaki, M., Inai, R., & Ramakrishna, S. (2005). Potential of nanofibermatrix as tissue-engineering scaffolds. *Tissue Engineering, 11*(12), 101-109.

Markovic, D., Karadzic, I., Jokanovic, V., Vukovic, A., & Vucic, V. (2016). Biological aspects of application of nanomaterials in tissue engineering. *Chem. Ind. Chem. Eng. Q., 22* (2), 145−153.

Miller, D., Thapa, A., Haberstroh, K., & Webster, T. (2004). Endothelialand vascular smooth muscle cell function on poly (lactic- co-glycolic acid) with nano-structured surface features. *Biomaterials, 25*(1), 53-61.

Mohamed, A., & Xing, M. (2012). Nanomaterials and nanotechnology for skin tissue engineering. *Int J Burns Trauma, 2*(1), 29–41.

Rivron, N., Liu, J., Rouwkema, J., de Boer, J., & van Blitterswijk, C. (2008). Engineering vascularised tissues in vitro. *Eur Cell Mater., 15*, 27-40.

Schwartz, M., & DeSimone, D. (2008). Cell adhesion receptors in mechanotransduction. *Curr Opin Cell Biol., 20*(5), 551-556.

Tuzlakoglu, K., Bolgen, N., & Salgado, A. (2005). Nano- and micro-fiber combined scaffolds: a new architecture for bone tissue engineering. *J Mater Sci Mater Med., 16*(12), 1099-1104.

Vogt, A., Wrobel, G., & Meyer, W. (2005). Synaptic plasticity in micropatterned neuronal networks. *Biomaterials, 26*(15), 2549-2557.

Wang, J., Apeldoorn, A., & Groot, K. (2005). Electrolytic depositionof calcium phosphate/chitosan coating on titaniumalloy: growth kinetics and influence of current density,acetic acid, and chitosan. *J Biomed Mater Res A., 76*(3), 503-511.

Wen, X., Shi , D., & Zhang, N. (2005). Applications of Nanotechnology in Tissue Engineering. In Nalwa, H. (eds.). *Handbook of nanostructured biomaterials and their applications in nanobiotechnology*. American Scientific Publishers.

Xie, J., Willerth, S. M., & Li, X. (2008). The differentiation of embryonicstem cells seeded on electrospun nanofibers into neurallineages. *Biomaterials, 30*, 354-362.

Xu, C., Inai, R., Kotaki, M., & Ramakrishna, S. (2004). Aligned biodegradable nanofibrous structure: a potential scaffold for blood vessel engineering. *Biomaterials, 25*(5), 877-886.

Yang, F., Murugan, R., Wang, S., & Ramakrishna, S. (2005). Electrospinning of nano/micro scale poly(L -lactic acid) aligned fibers and their potential in neural tissue engineering. *Biomaterials, 26*(15), 2603-2610.